CONCRETE WALKWAYS PROJECTS

by Jack King

THEODORE AUDEL & CO.

a division of

HOWARD W. SAMS & CO., INC.
4300 West 62nd Street
Indianapolis, Indiana 46268

FIRST EDITION

FIRST PRINTING—1974

International Standard Book Number 0-672-23804-7

Preface

Skyscrapers, factories, homes and buildings of every nature are built with concrete foundations. Millions of miles are driven on concrete highways each day, and 200 ton airliners and transport planes land on long concrete runways around the clock. Waste material is carried away through concrete pipes. Our water supply is piped through concrete and millions of miles of telephone and electric cables are housed in concrete pipes or tunnels. Bridges, dams and other large structures are also built of reinforced concrete.

In fact, the use of concrete is so ingrained in our daily use that we sometimes take it for granted. Even a simple thing like walking to the bus line would assume difficult proportions were it not for the flat surfaces we walk on. And imagine a bus going several miles across rough terrain. Concrete is one of our truly needed conveniences we would find it hard to do without.

There are special kinds of concrete developed for a variety of uses. Among these are: 1. reinforced concrete; 2. precast concrete; 3. prestressed concrete; 4. air-entrained concrete; 5. high early-strength concrete; 6. concrete masonry.

Although the total picture of the concrete industry is complex and the technical knowledge that goes into it is immense, this book will make the construction of a walk or driveway an easy task for the home owner. The secret of doing the job right is to study each move, step-by-step, and then do the simple jobs first until they are understood—you will enjoy the accomplishments of a concrete project built by yourself and you will save money.

Jack King

Contents

SECTION 1

Introduction

The most important reason for you to do your own home improvement jobs is probably the savings involved. Whether you do them for this reason or for the pleasure of accomplishment, the following pages of instructions and pictures will help you do a professional job of building with concrete. You will also learn to make attractive stepping stones and find numerous other hints that will help you work with concrete. Since the instructions start with the very basic things like practicing with sand, you may wish (if you already have some experience with concrete) to skip this section and go on to the work of making stepping stones or to the walk itself. Very little technical information is included as very little is needed to do a good job. One thing you should keep in mind is the difference between concrete and cement. When we speak of cement, we are referring to the powdered dust which comes in bags; and concrete refers to the mixture of gravel, sand, cement and water.

This book will also prepare you for many of the unexpected problems you may encounter when doing these small projects. It will also inform you of the various tools needed for a particular project and explain their proper usage; the proper size lumber to be used in making forms; and a step-by-step procedure you can follow when building sidewalks, driveways, steps, and patios, plus the proper procedure to follow when repairing old damaged concrete.

SECTION

SECTION

Tools and Equipment

This is a layout of most basic tools you will need to make stepping stones, build forms, and build a sidewalk. Any additional tools will be mentioned further on. Some of these shown, such as the electric power saw, makes the work easier but are not absolutely necessary. The bull float (the tool with the long handle on the right) makes finishing concrete walks much easier, but you can get by with just the darby (the long handled float on the left). Rather than buy some of the more expensive tools, it is suggested that you rent these.

The tools shown in Fig. 1 are:
Darby,
Jointers (2),
Soft-bristled broom,
"Come along" (cement rake),
Wire mesh cutters,
Chalk line & bottle of chalk,
Roll of line (string),
Tape measure or folding rule,
Electric power saw with extension cord,

Trowel,
Aluminum or magnesium float,
Wrecking bar,
Claw hammer,
Bull float (aluminum),
Level,
Framing square,
Rubber boots & gloves,
Sledge hammer,
Panama shovel (and a flat shovel),
Handsaw,
Pick,
Saw horse.

SECTION

SECTION

Building a Form for Stepping Stones

Step 1. Cut the lumber. Cut two pieces of 2 × 2 lumber approximately 8 feet long, as shown in Fig. 2. Cut six pieces of the same type lumber 16 inches long.

Step 2. Fit the pieces together. Measure off three 16 inch sections on the long boards. (See Fig. 3.) Leave about

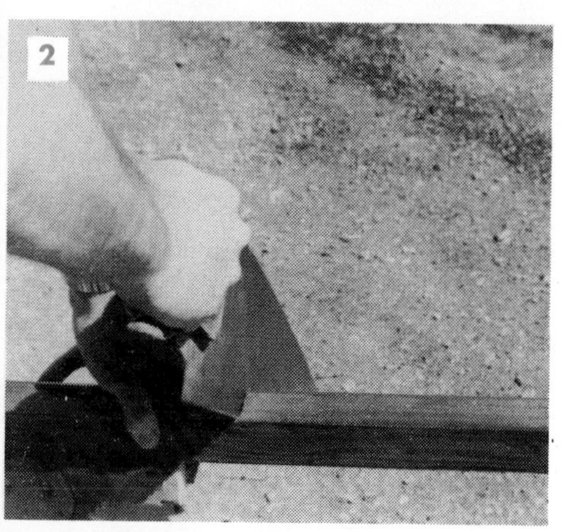

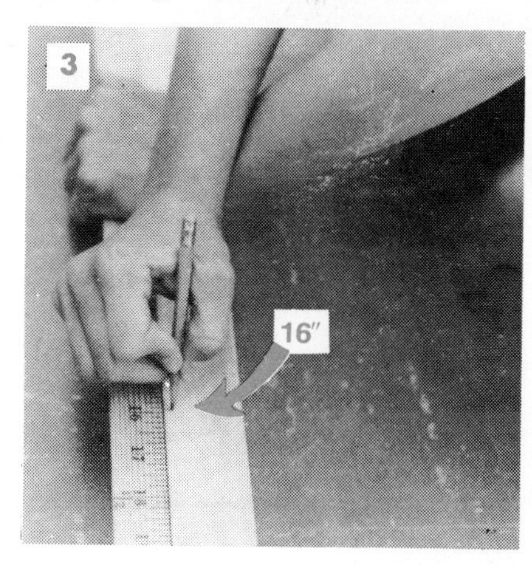

10 inches or so between each section, and fit the 16 inch pieces between the long pieces forming three 16 × 16 inch squares. See Fig. 4. Lay off both long boards together, as shown in Fig. 4, so that the small spacers will form perfect squares with the long boards. If you would rather do one stone at a time, you may make just one 16 × 16 inch square, but if you want to make three stepping stones at a time, do as shown in Fig. 5.

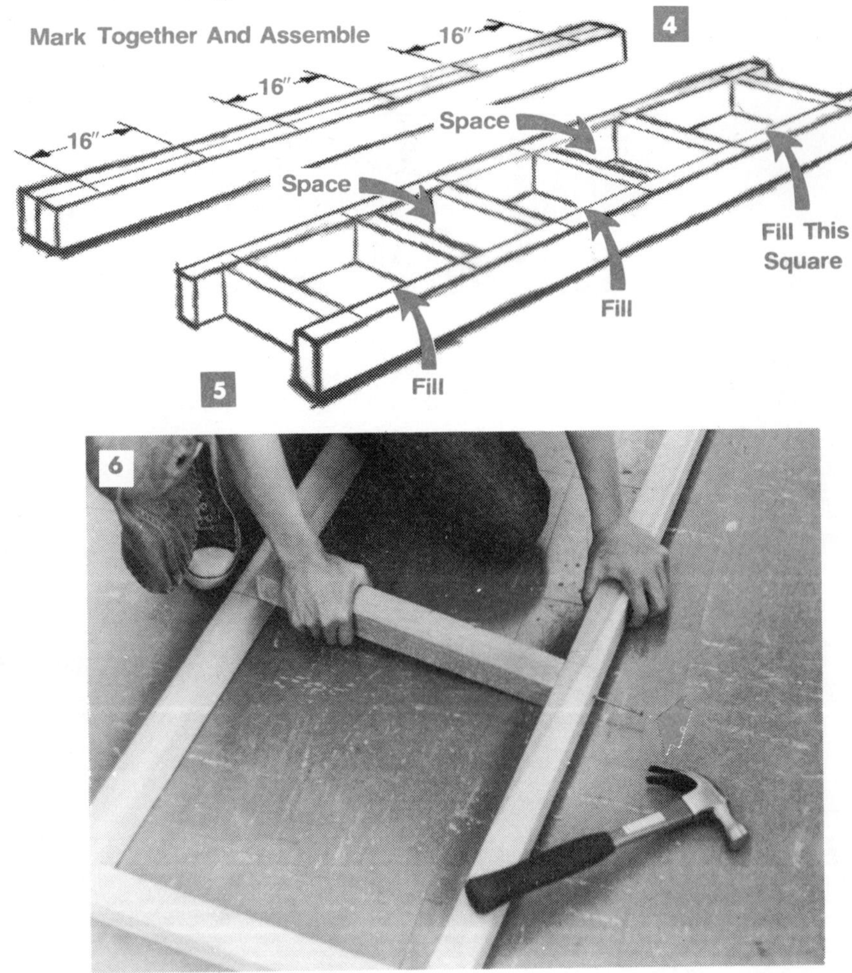

Step 3. Nail the form together. Nail the short pieces (Fig. 6) between the long ones where marked. Either leave the nail sticking out ¼ inch or use scaffold nails for easy disassembly of the forms after the concrete has hardened. If you are careful in taking them apart, you can use them several times. See Figs. 7, 8 and 9.

7

8

9

Step 4. Squaring the form. After the form is assembled, square it with a framing square, and place a short brace across one corner to hold it in position. See Fig. 10. If you plan to make your stepping stones on the ground near the place you will use them, a better way to secure the form is by driving small stakes in the ground on two sides of the form. See Figs. 11 and 12.

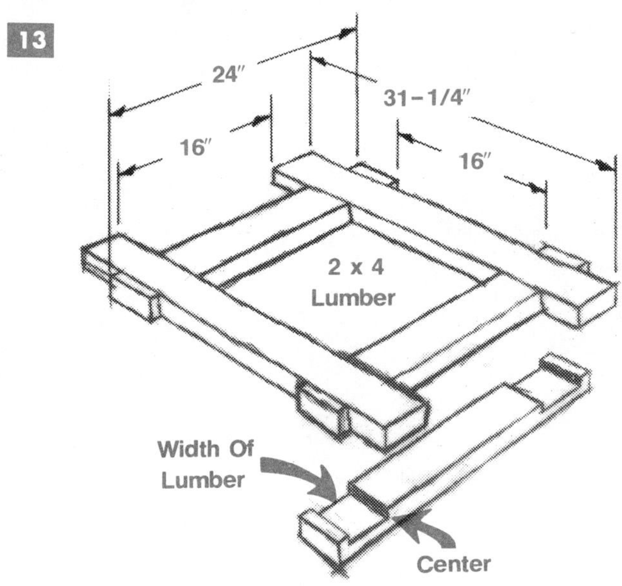

If you want to be more elaborate in making your forms, you can cut your boards as shown in Fig. 13. Cut your 2 × 4's (for more strength) 31¼ inches in length. Measure in 4 inches from each end and cut notches to the width of the form lumber used. If 2 × 4 lumber is used, the notch would normally be 3⅝ inches wide. The notch depth would be cut down ½ the thickness of the form lumber. Assembled as shown in the illustration (Fig. 13) the forms can be put together, held in place, and knocked-down without the use of nails. With this design, the forms can be used many times without wear or destruction to the form material.

PRACTICING WITH SAND

Step 1. Fill the squares with sand and screed. Fill up all three squares of the form with damp sand, as shown in Fig. 14. Pile it slightly higher than the form sides. Just as if the sand were concrete, use a small straight board to "screed" (rake) off the sand level with the form sides, as shown in Fig. 15. Saw back and forth as you move across the square. If the sand piles up too much so that it becomes difficult to screed, rake some of the excess off and resume "sawing." Move the straight edge in the direction of the arrows (Fig. 15) as you move across the surface. If small holes or depressions are left in the surface behind the screed, sprinkle sand there and go over it with the screed.

14

Step 2. Practice edging. After you have raked the sand flat in your form, take the edger and position it so that the curved edge will round off the edges of the practice sand block. Move it carefully around the perimeter keeping next to the form and keeping the flat part flat against the sand. Practice this several times until you are satisfied that you could edge a concrete block as well. See Fig. 16.

Step 3. Practice cutting joints. Mark the middle of your form with a pencil, as shown in Fig. 17, and a nail to easily find the mark. When you pour concrete, the screeding will cover the pencil marks (as you will see). Fit the jointer or groover (called either) so that the groove will line up with your mark. Adjust the straight edge accordingly. See Fig. 18. Although you will not need joints in your stepping stones, this is a good opportunity to practice cutting joints. When you need to cut joints in your sidewalk, you will be thoroughly familiar with this operation. Joints not only give your walk esthetic value; they provide a breaking point in case of cracking tendencies caused by shrinkage when drying out or cold weather freeze.

Step 4. Practice with the float. No illustration of the floating operation is shown. The main thing to remember

17

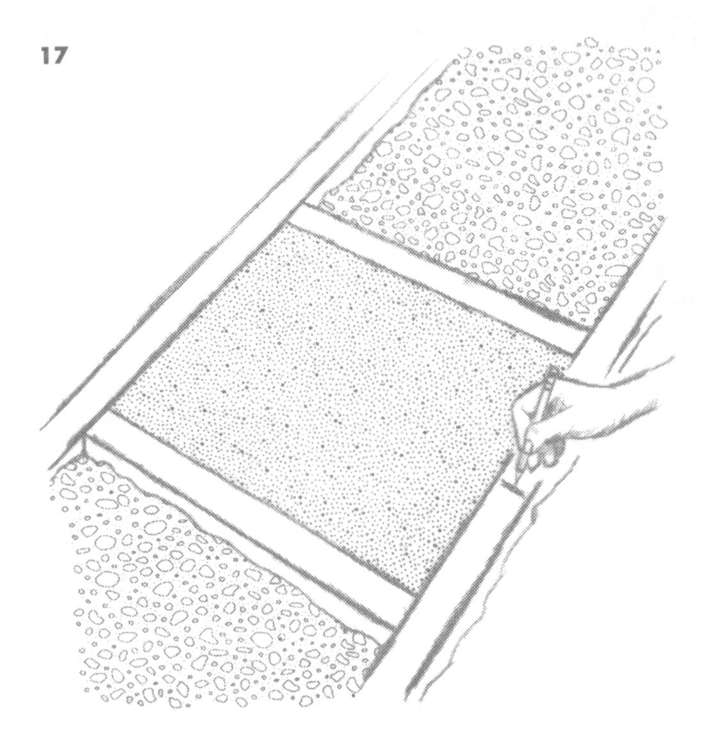

18

when using the float is to keep it flat on the surface and move it around in small circles. Do not overdo the floating on real concrete, and do not waste too much effort on this sand floating. Although it will be helpful to know how to hold the float, the sand will not work quite the same. When you are making stepping stones, you will get more valuable experience floating real concrete.

Step 5. Practice with the trowel. The trowel will smooth the sand about the same way it will smooth the concrete, with the exception that there will be no real fine particles of cement to work to the surface. Place the trowel flat on the sand surface at one edge of the form and lifting the leading edge slightly, move it in a smooth arc to the other side. Lift the opposite edge and return in the same manner to the other side. This is the most important thing to learn to smooth the surface carefully. Do not damage the round edges you have made with the edger. Dump the sand from the form, and practice the whole operation as many times as you feel it is necessary. See Fig. 19.

MAKING STEPPING STONES

Step 1. Figuring materials. To make stepping stones, it will be much easier to buy 50 or 90 pound bags of premixed concrete. The cubic foot measure is given on the bag. To figure how much you need, just measure roughly the form you are going to use. In this case, using the form in Fig. 14, 2″ × 16″ × 16″ equals 512 cubic inches. Since there are 1728 cubic inches in one cubic foot, and a 90 pound bag of premixed concrete equals about ⅔ of a cubic foot, each bag will make two stepping stones for this form.

Mixing your own concrete

If you are going to make a large quantity of stepping stones, and you want to save money by mixing your own concrete, a good mixture to use is 3 parts gravel, 2 parts sand, and 1 part cement and just enough water to make the concrete easy to work. The less water you use, the stronger will be your concrete. Stir the mixture dry first before adding water.

Later, when you plan to build a sidewalk, you may decide that you can save money by mixing your own concrete as you did for your stepping stones. While it is possible to save a little on the right size jobs, it is advisable that you order concrete from a ready-mix company. Without much experience in concrete work, you would find that additional labor and equipment rental required would not save you that much, and would increase the details that you must think about. It is not practical for a beginner to mix his own. For stepping stones you can mix your own, if you wish, but for a sidewalk order *ready-mix.* In some cases, there is an additional delivery charge for *ready-mix* when less than 5 yards are ordered.

Step 2. Tools needed. After you have obtained a float, edger, groover (jointer) trowel, straight edge, hammer, and a small plastic or metal tub to mix your concrete, pour enough of the premixed concrete into the container to make at least one stepping stone. See Fig. 20.

20

Step 3. Add water. Add water slowly as you mix until the mixture is wet throughout but still stiff. See Fig. 21. If you get it too runny, add more dry mix.

Step 4. Fill the form with concrete. Scoop the well-mixed concrete into your form and tamp it down into corners and edges with a short stick or the end of your float. Tap on the sides of the form with a hammer. This operation (called puddling) will make the concrete settle against the form and prevent honeycombs (small holes). See Figs. 22 and 23. Painting the inside of the forms with paraffin oil before pouring the concrete will keep the concrete from sticking to the wood. This is not absolutely necessary, but it helps keep the forms clean for re-use. If you do not use oil, wet the inside of the forms with water.

Step 5. Screed the concrete. As you tamp the concrete into the form, pile it slightly higher than the sides and "screed" the concrete with a straight edged board. Move

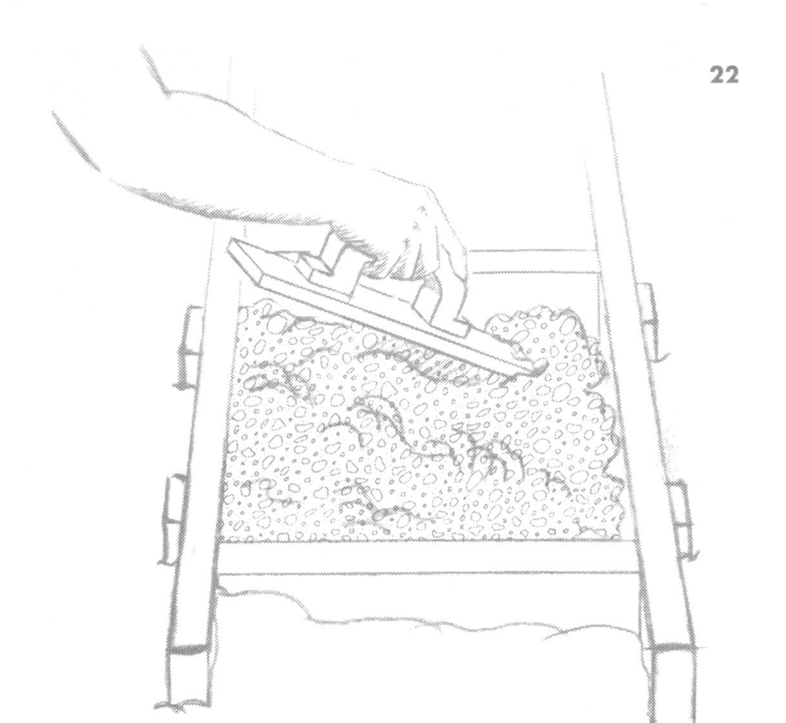
22

23

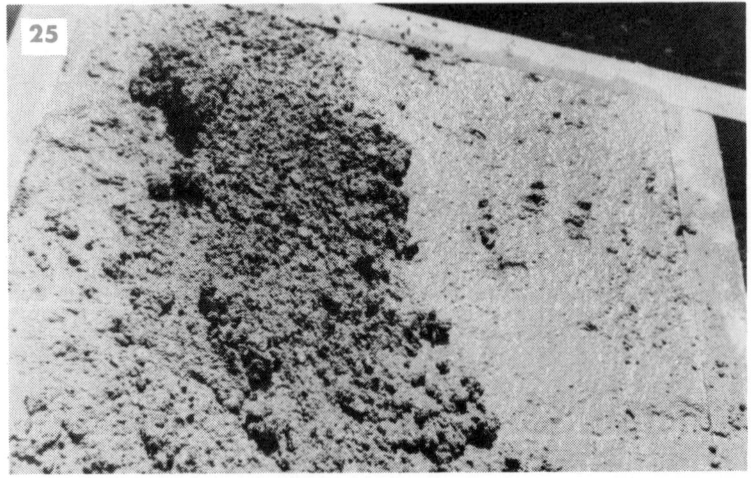

the board from end to end (arrows) back and forth as you pull the concrete across the square, raking the excess from sides. See Fig. 24. If you leave small surface holes or depressions, sprinkle some of the excess concrete into the holes and go over the top with the straight edge again. See Fig. 25.

Step 6. Floating the concrete. Float the concrete once, immediately, by placing the float flat on the surface and lightly move it in circles (See Figs. 26 and 27). After this effect is achieved, let some of the water evaporate before working it more. Too much floating when water is present

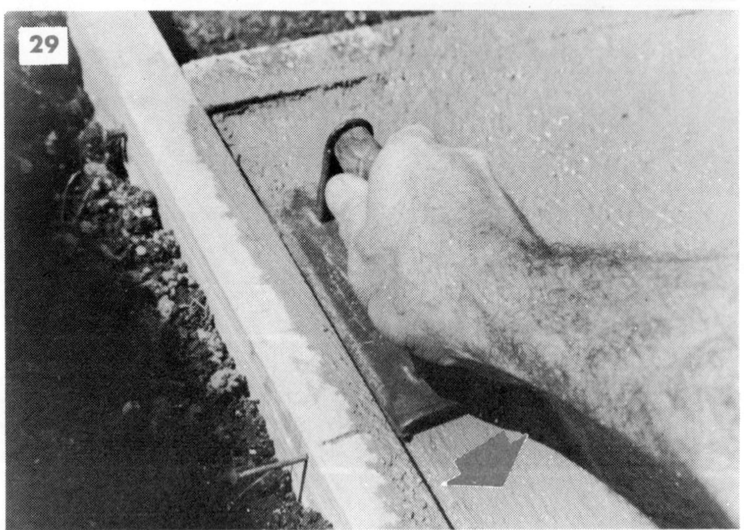

might cause a weakness in the surface which will flake off or crack later, as shown in Fig. 28. The main purpose of floating is to imbed the larger aggregates beneath the surface and bring the smaller particles to the top for better working ability.

Step 7. Edging. After the concrete begins to dry and stiffen, (you can tell by a slight color change) place the edger along the form and move it along the wood so that the rough edge is curved all around the perimeter of the stone. See Fig. 29. This will leave a slight line where the inside of the tool travels, as shown in Fig. 30.

On a sidewalk, the border gives a neat finished look, but on a stepping stone, you will probably prefer that the line is troweled over. If you damage a corner while edging, just scrape some of the paste-like cement off of the form, place it in the damaged spot, and go over it with the edger again.

Step 8. Cutting joints or grooves. On a stepping stone, you will not need joints, but this is a good opportunity to practice on a smaller scale that operation you will need on your sidewalk later. This job, if done correctly, will do much to make your sidewalk look like a professional job. After you finish a practice joint on the stone, fill it in with excess and float it over again. You will notice that the pencil line indicating where the joint begins is also marked by a nail, as shown in Fig. 31. The concrete spilling over the form will

hide a pencil mark, but you can find the nail more easily. If you do this before the concrete is poured, you will not have to stop and measure for joints when you are busy finishing. Place the sharp edge into the concrete on both sides of the form and adjust the straight edge accordingly.

Sometime before you build your sidewalk, it would be a good idea to observe one or several new sidewalks to see the finished joints and edges left by the builder. When putting the finishing touches on a walk, it looks better to use the edger one last time then run over the joints with the jointer, letting the tool cut into the edger marks but stopping short of the walk edges as viewed in Fig. 32.

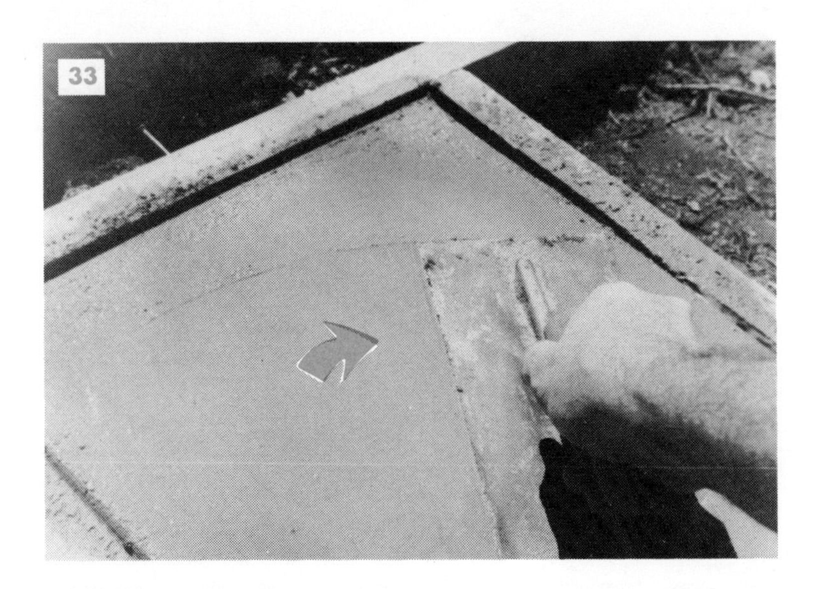

Finishing the stepping stones

Fill in the practice joint you made in the stone with excess concrete and float the surface again. Use the edger again, this time floating out the lines left by the tool. As the concrete begins to harden, the edger will leave a smoother finish. This is a good indication that it is time to use the trowel.

Step 1. Trowel the stepping stone. Place the trowel flat on the surface and move it in a smooth arc lifting the leading edge slightly, as shown in Fig. 33. If your trowel leaves a "washboard" surface, either you are lifting the leading edge too high or it is too wet to trowel. Flatten the trowel and make another test arc. If you still make ridges, float out the surface and wait until the trowel leaves a smooth surface. If the top looks like the surface shown in Fig. 34 when you go across it, you are doing fine. Go over it until the entire top is smooth.

Step 2. Swirl finish. To obtain a swirl design, let the slick finish set for a short while; then set the trowel flat on the surface and move it around in small circles, lifting straight up as you finish a run across the top. If you are careful, you will not cut into the top. So far, you have done three types

of finishes, any of which you may leave for a nice looking stepping stone. The usual finish for a sidewalk will be the broom finish. You may want to try this first on one of your stepping stones. See Fig. 35.

Step 3. Broom finish. Simply set the soft-bristles of the broom down on one side of the stone and drag carefully across the surface. A variation of this is the wavy broom finish. Do the same thing but make large S turns as you bring the broom across. Start at a slight angle rather than perpendicular to the side and make the surface ridges wavy. See Figs. 36 and 37.

Step 4. Removing the forms. After the concrete has hardened, preferably the next day, take the forms apart carefully, scrape them, apply more paraffin oil (if you are using it) and get them ready for new stepping stones. You may remove forms carefully a few hours after pouring, but it is well to have some means to protect the edges for about twenty four hours. See Fig. 38.

Step 5. Curing. Water the new stepping stones several times in the next few days, especially if the temperature is 70 degrees or more. This will keep the concrete from drying out too fast. It will still get hard even with the water on it. Do this for several days.

Other shapes and types of stepping stones

Step 1. For a rustic looking stone. Place a small pile of concrete equal to the amount used to make the other stones on the ground. Tamp it down flat to about 2 inch thickness, and work the float in a circular motion. Finish with a rough look, as shown in Fig. 39, or:

Step 2. Rustic finish with a paint roller. Roll an old paint roller across the surface. See Fig. 40.

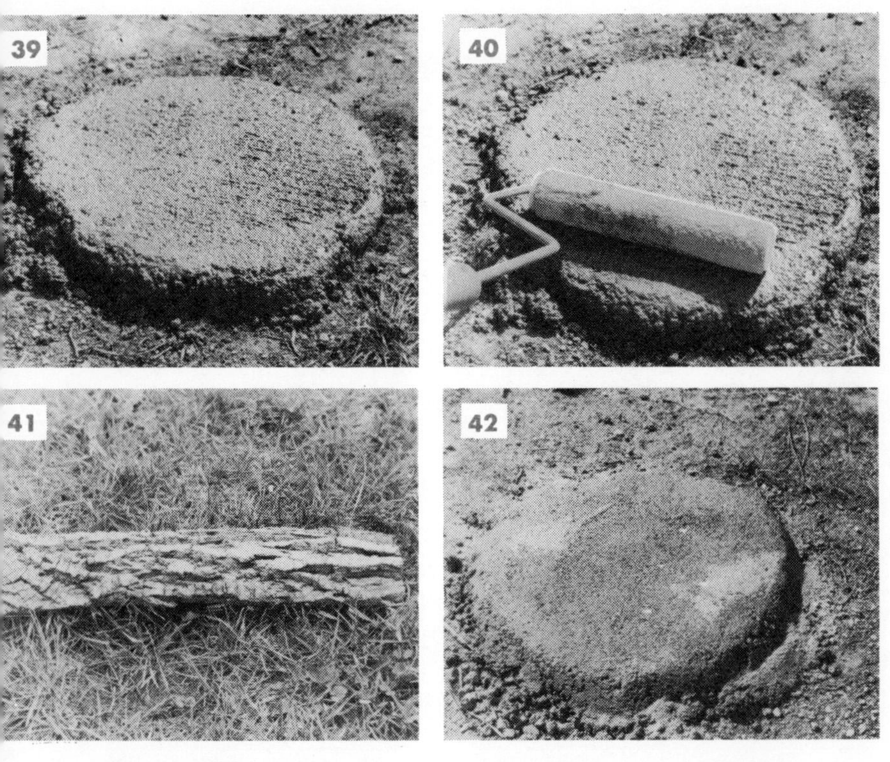

Step 3. Rustic log finish. You might obtain a rustic looking finish by rolling a rough bark log over the surface. This stone was done with the small piece of bark shown. A bigger one like this might be more effective—depending upon what you want. See Fig. 41.

Step 4. Sand or gravel top. Another method to achieve a somewhat rustic finish is to pile sand or washed gravel on the surface, tamp it down and float it some. Then later wash the loose sand or gravel off with a hose. See Fig. 42.

Step 5. Other shapes. You can make a stepping stone any shape you build a form. Curved edges can be achieved by curving a piece of masonite or scored plywood, holding it in place with stakes, or you might use old barrel hoops. See Fig. 43. Place the smaller opening to the top so that it can be lifted off when the concrete sets up. Almost any

43

Bending Plywood For A Curved Form

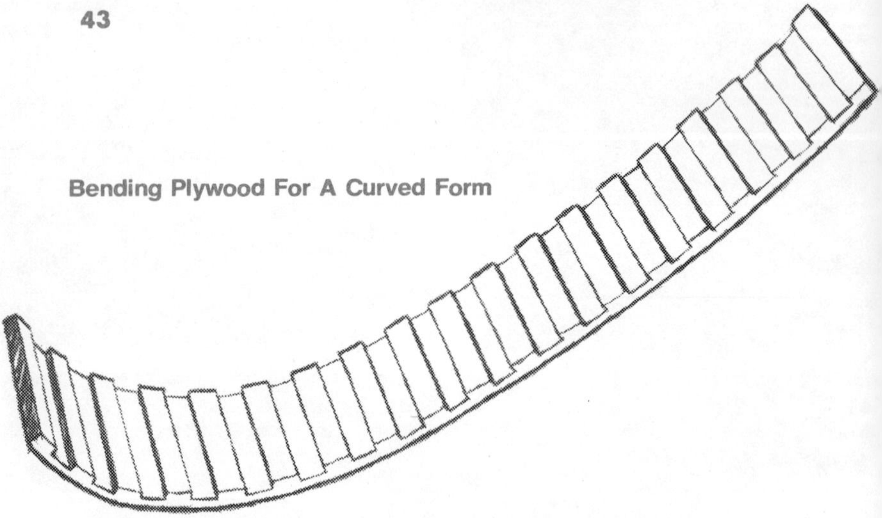

shape or finish is acceptable as long as the stepping stone does not become a "stumbling block" (i.e., does not contain a hole that a heel might catch in).

SECTION

SECTION

Building a Sidewalk

Step 1. The planning stage. While no elaborate blueprint is needed for a sidewalk, it may help to look over the area to determine the best method or shape for the walk, and then make a simple sketch. These two sample sketches (Figs. 44 and 45) will give you some idea of how to plan your walk and figure the amount of concrete needed

By using graph paper for your plan, you can easily count the squares and figure how many square feet your walk covers. Use one square for each square foot. Once you know the square footage of the walk, you can figure cubic measure in this manner:

> If the walk is four inches thick (a normal walk), divide the square feet by 3. Since there are 27 cubic feet to a cubic yard, divide the cubic feet in the walk by 27. Read the sample problems. If your walk is six inches thick, you would divide the square feet by 2, and then divide this by 27. If you have less than one cubic yard, you should order one yard. The cost will be about the same.

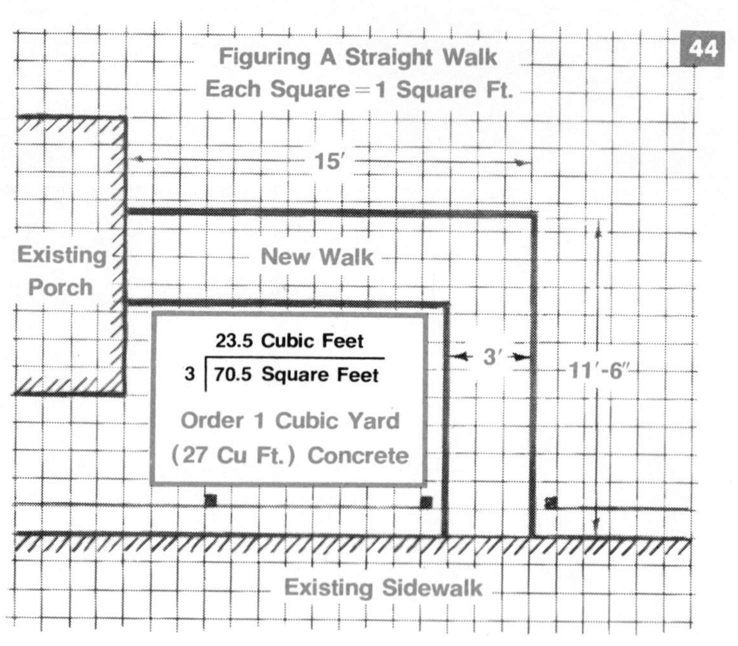
44
Figuring A Straight Walk
Each Square = 1 Square Ft.
15′
Existing Porch
New Walk
23.5 Cubic Feet
3 | 70.5 Square Feet
Order 1 Cubic Yard
(27 Cu Ft.) Concrete
3′
11′-6″
Existing Sidewalk

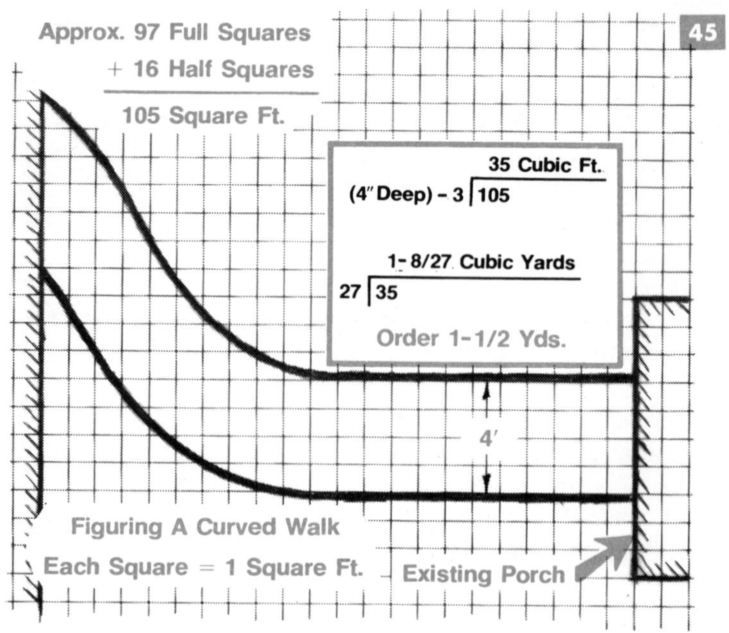
45
Approx. 97 Full Squares
+ 16 Half Squares
105 Square Ft.
35 Cubic Ft.
(4″ Deep) – 3 | 105
1- 8/27 Cubic Yards
27 | 35
Order 1-1/2 Yds.
4′
Figuring A Curved Walk
Each Square = 1 Square Ft.
Existing Porch

Step 2. Choosing form lumber. Pick out some straight 2 × 4's for the walk form. See Figs. 46 and 47. If when holding the 2 × 4 flat side up, the board curves up or down slightly, you might still be able to take the curve out by driving stakes in the ground. If the board curves up or

down when holding the narrow side up, reject it and select another board. The ideal one has no curves. Ask for building grade 2 × 4's unless you have an important use for the boards after you build the form. If the 2 × 4 is cleaned immediately after use, it is still in good condition. The building grade boards are cheaper. Some lumber yards may have other names for this grade lumber such as number 2. You will also need 6d, 12d or 16d common nails. Also some scrap wood for stakes and splices.

Step 3. Locating lines and establishing grades. The question of where to place the lines for your walk contains several factors for you to consider. Read the following steps and study Fig. 48 for establishing your sidewalk at a square angle (perpendicular) to any existing structure (as a porch). Then consider drainage. Find the highest point of the walk and figure how much to allow for drainage. The following photos and instruction should simplify this.

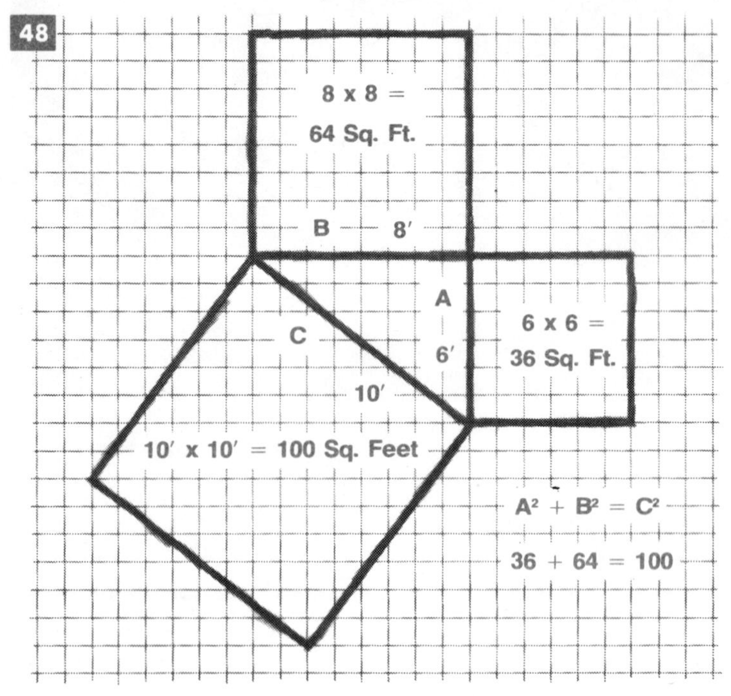

Step 3a. Reference point. Find a reference point from which to start your measuring. Where one edge of the walk will meet the souse, make a mark and lay a 2 × 4 flat on the ground extending to the approximate place for a 90 degree angle. See Fig. 49. From the reference point, measure 6 feet along the house or porch.

Step 3b. 6-8-10 Method. Measure 8 feet along the 2 × 4 from the reference point. This will give you two legs of a triangle, one 6 feet, and one 8 feet. See Fig. 50.

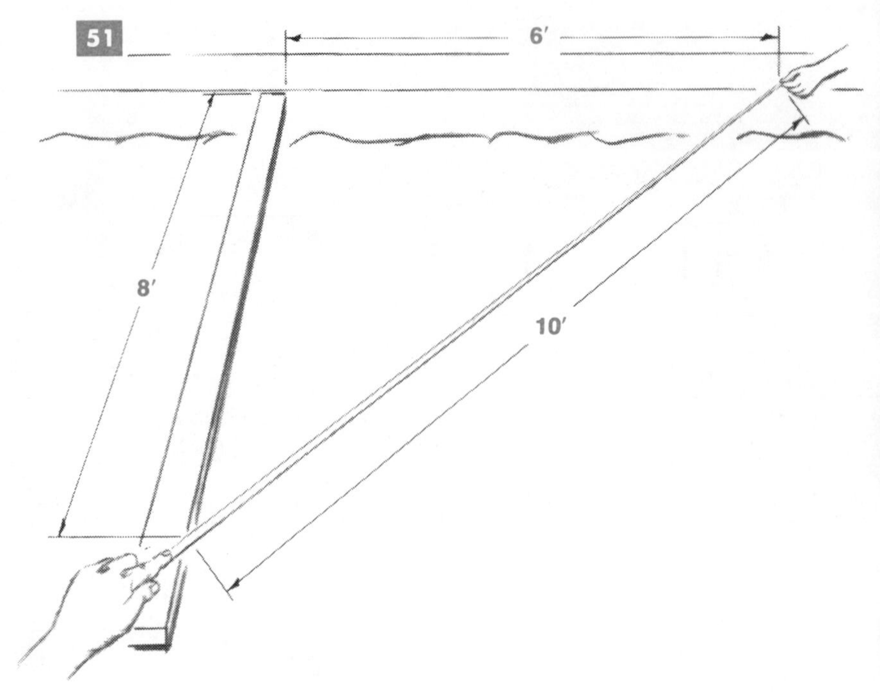

Step 3c. Complete the triangle. Have someone hold the tape at the 6 foot mark on the house. Hold the 10 foot mark on the tape so that it meets the 8 foot mark on the 2 × 4. See Fig. 51. You will have to adjust the 2 × 4 some so that both measurements meet. Set two small stakes to mark this 90 degree angle. See Fig. 54.This is based on the geometric formula by Pythagoras—"The sum of the two squares on the sides of a right triangle is equal to the square on the hypotenuse (the long side.)) Any multiple of 3-4-5 will work (6-8-10), (9-12-15, etc).

Step 4. The line. With the two stakes as a new reference point to indicate a 90 degree angle with the house, set the form to the correct height. If the walk is not connected to a structure, simply set up one line for an edge of the walk. Here, iron stakes are used, but wood is just as good. See Fig. 53. If the ground is high, dig a trench with the pick so that the 2 × 4 top will be on line. See Fig. 52. You can also place

52
53
54
Angle Desired

a nail on the inside edge of the 2 × 4, stretch the string to another stake, and adjust the 2 × 4 until the inside edge is right on the string, as shown in Fig. 55. Flip the string to clear it, as shown in Fig. 56. Additional lengths of 2 × 4's are connected with splices.

Step 5. Setting stakes. When the 2 × 4 form is close to the point you want it, drive stakes about every 5 or 6 feet. Nail the 2 × 4 at the string height. If the stake holds the form too far one way or the other, hit the ground near the stake to force it over. See Fig. 57. If you hit the stake itself, you will either knock it loose from the 2 × 4 or loosen it so it is not firm in the ground.

Step 6. Spacer. When one side of the form is set to your satisfaction, cut a board to the same length as the width of the walk. Use it to set the other side the exact distance from the side already set. See Fig. 58.

Step 7. The height of the other side. If the first side of the form is set properly and you want a level walk, just level it across with your level as you set the other stakes to the right height. If you want some fall for drainage to the other side, place a small block of wood under the level for the low side. See Fig. 59.

Some rules to follow:

1. If the walk starts at the porch, give it a slight amount of fall to drain away from the house.

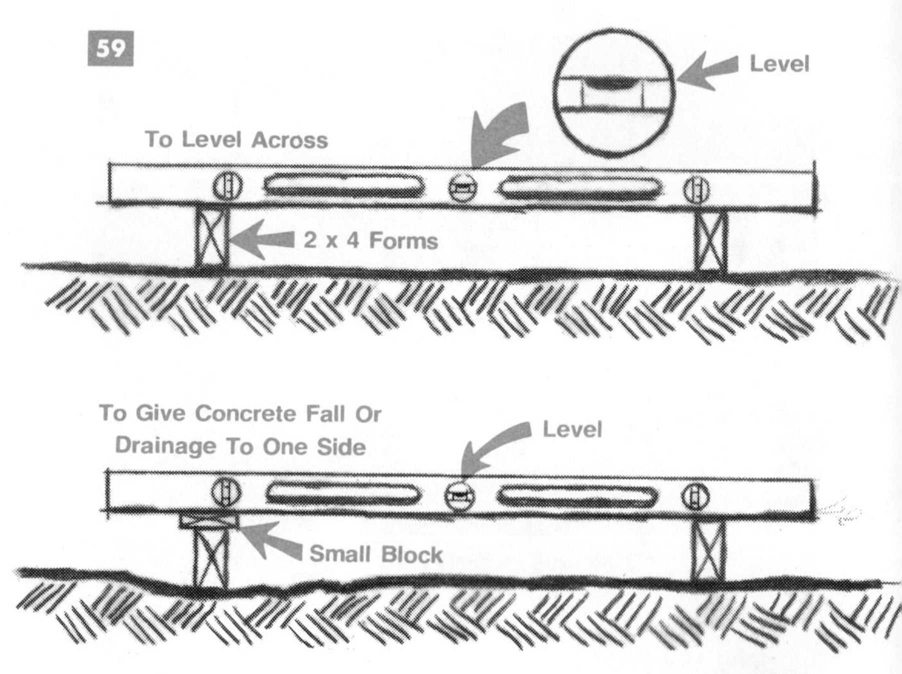
59
Level
To Level Across
2 x 4 Forms
To Give Concrete Fall Or
Drainage To One Side
Level
Small Block

60

2. If the walk is parallel and close to the house, let it fall slightly to the outside, away from the house.
3. If the walk follows the contour of the yard, you can let the drainage go the best route.
4. A flat narrow walk will not keep much water trapped on top, so leave it flat, if you wish.
5. Avoid the situation that allows the fall to go too low at one end without an outlet for water, as shown in Fig. 60.

Step 8. Grade the sand or dirt between the forms. Using a flat shovel, or pick if necessary, remove all sod from between forms and grade it level with the bottoms of the 2 × 4. See Fig. 61. If there are low places, use sand for fill, as shown in Fig. 62. Too many low places will add to the amount of concrete you will need. A small sand rake can be made by nailing a board onto a 2 × 4, as shown in Fig. 62. Sand can be tamped with a flat square board nailed to a handle, or if you have a lot of tamping to do, you can rent a tamper with a motor. See Fig. 63.

62B

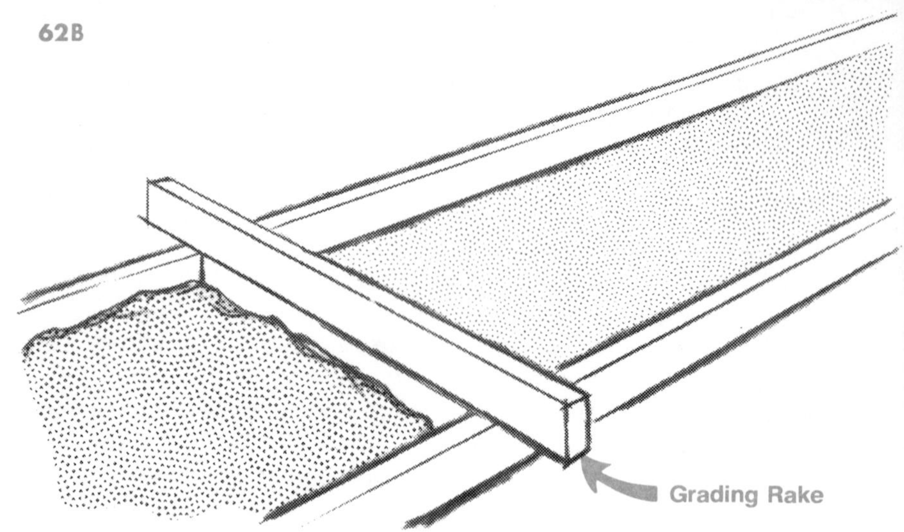

Step 9. Reinforcing. For a simple walk, you will not need reinforcing. You can use it if you want a "super strong" walk. If cars will run over certain parts of your walk, use wire mesh. When using mesh, as you pour the concrete, lift the wire with a wrecking bar so that it rests in the lower third of the concrete. You can lift it with your hand also. Wear leather gloves as the ends are very sharp. *A safety note:* When unrolling wire mesh, be sure the person holding down the other end does not step off. The wire will try to roll back up, and in so doing, could deliver a very bad injury or blow. Cut it with heavy wire cutters; then two people can turn it over into the form. It will not unroll when put in place, but it will bulge up and might have to be bent a little to stay in position.

Step 10. Expansion joints. An expansion joint is composed of material that will give a little when freezing and thawing pressures are exerted upon the concrete. In a small walk, one could be used where the walk meets the house, and

about every 25 or 30 feet. There are two good ways to place these expansion joints. 1. If you pour half of the walk one day and half another, use the stopping point to place a joint. See Fig. 64. After you take out the spacer used to stop the concrete, place the joint against the edge of new concrete. When you pour again, shovel some wet concrete against the joint. Let it stick above the surface, and cut it off with a knife the next day. 2. If you do the walk in one pour, place a spacer with the joint against it; pour on each side of the spacer; and when the concrete is holding the joint firm without the 2 × 4 to hold it straight, dig out the 2 × 4 and fill the hole with excess concrete. See Figs. 65 and 66. This is one reason to have the truck driver leave you a small extra pile of concrete.

Step 11. Mark forms for all joints. Go back to the section on cutting joints for stepping stones, and follow instructions for marking the form with nails. Although most pros do not worry about this formality, a beginner should figure in advance where the joints will be cut. One way is to measure the length and divide it into pleasant looking squares or rectangles. If your walk is four feet wide, you could make

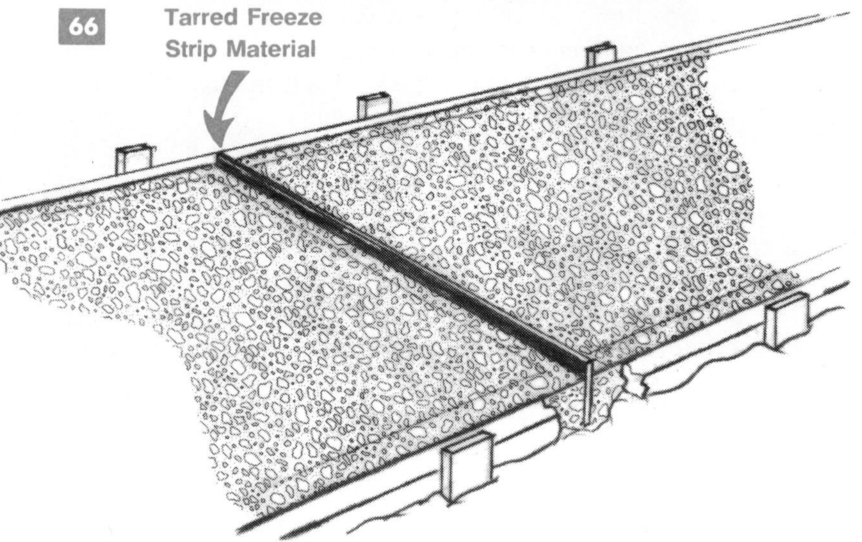

your joints every four feet, working out the difference with a smaller square near one end. This is not a major problem, but advance thinking about it will help you end up with a good looking walk.

ORDERING THE CONCRETE

The easiest way for a person pouring their first concrete job is to order *ready-mix* from a well-known concrete firm (the nearest one to you is not always the best, although most are reliable) and tell the dispatcher how many yards you need at what time. Do not always count on getting it at the time specified, although dependability is one of the concrete company's stocks in trade. Check the weather forecast the night before. If it seems favorable, put the order on "a will call basis," that is, you will verify it the next morning.

If the next day has a cloudless sky, and it is fairly cool, it could turn out to be the right day to pour. It is hard to guarantee a perfect day (as you well know) so take a chance and order the concrete. If the good day follows a few weeks of rainy weather, you will find that all of the big contractors are pouring concrete that day also, and you will have to be worked in between these big orders. This is not to discourage you but just prepare you to expect possible delays.

If you do not know how much concrete you need when you call the night before, tell the dispatcher what the size of your forms is, and what you are pouring, and he can tell you what you need. Concrete companies are among the best companies to do business with; you will find them concerned to make an effort to deliver on time. And if you should get bogged down with work, not one of them will pile the concrete on the ground and leave you stranded. A seasoned driver knows a beginner when he sees one and will go all out to be of assistance.

SECTION

SECTION 5

Pouring and Finishing

Step 1. The concrete arrives on the job. You should be at the site of your concrete job about an hour before the concrete is due to arrive. This will give you time to re-check measurements, make sure all of your equipment is there, and see that your helpers (if any) have arrived. You could probably handle a three foot wide walk by yourself, but if it is bigger, you will need someone to help you rake the concrete down (screed). The hardest work is while the concrete truck is there — while you are placing the concrete and screeding it. After this, you can take it easier as you work. If the concrete builds up in the chute and does not pour readily, ask the driver to add a little water. You can trust his judgment as to how much. See Fig. 67.

67

After it flows from the chute (as shown in Fig. 68) rake and shovel it into place. The "come-a-long" tool is good for this. The driver can place most of it where you need it by manipulating the chute. See Fig. 69.

Step 2. Screeding. If your new concrete begins at a pre-existing walk or from a porch, you will find it easier to start screeding by floating along the beginning edge. Place your straight edge (2 × 4) across the walk, and sawing back and forth, rake the concrete even with the top of the form. If you have another helper, he can pull down the excess with a come-a-long or a flat edge shovel. See Fig. 70. If you leave holes or depressions (or humps) in the surface behind the screed, sprinkle excess concrete in the holes and go over it again. Notice that the bull float is employed immediately behind the screed, as shown in Fig. 71. If you are working with limited help, you can screed some, and bull float the first portion; then screed some more.

Step 3. Bull floating. Place the bull float on one side of the walk, lower the handle which lifts the leading edge, push across to the other side, lift the handle with the edge toward you and pull it back across. A few times across (Fig. 72) and you will have the knack. Just go across two or three times

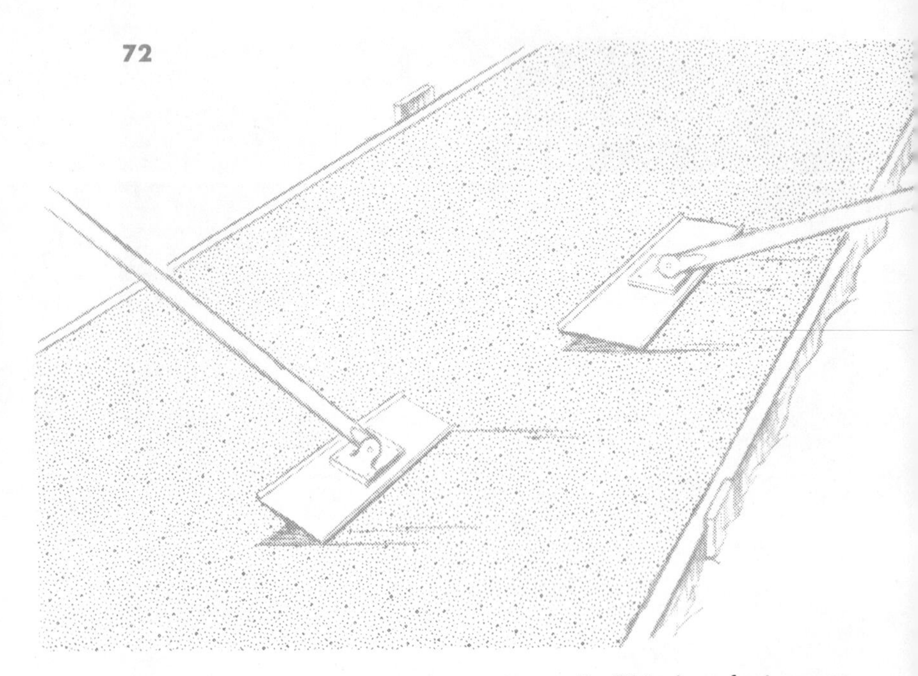
72

at each point and move down the walk. This is a fast operation as the large bull float covers quite a bit of surface at one time. If you are constructing a small walk, a darby will suffice. This long float-like tool is used just like you use the hand float only moving in larger arcs across the surface. The bull-floated surface will have a fine textured and level appearance when you complete this operation. See Fig. 73. If a portion, as on the walk end, has a slight slope and the concrete flows downhill, you can use the bull float to push the excess back up the slope and work it into the level part. See Fig. 74. If there is an excess that will not work smoothly into the level part, screed it at that point. Do not do any more to the surface at this time until the surface wetness appears to be disappearing. It will change color slightly. The concrete walk shown was poured on a day with the temperature about 40 degrees. The cool weather slowed the drying out process, and it was about an hour before the edging and floating was done. You may need to do yours sooner.

73

74

Step 4. Preliminary edging. Do not edge immediately. Wait until the concrete begins to harden at the edges, and then run a preliminary edge all around your form. You will do this again later, so do not be painstaking on it at this time. See Fig. 75.

Step 5. Floating. Making sure again that the surface is not too wet, move the float in wide sweeping arcs flat on the surface to bring the finer particles to the top, as shown in Fig. 76. Use an aluminum or magnesium float. Work one side of the walk at a time moving the float to about the center.

Step 6. Put on the broom finish. With the preliminary edging done and a careful floating job completed, wait until the top dries out a little more. When you can pull a soft-bristled broom across and leave a fine-textured broom finish, go over the surface very lightly once, overlaping each time. If you pull too many little rocks or chunks of surface up with the broom, then wet the broom. However, if the broom seems to sink into the surface, wait until the concrete dries some more. See Fig. 77.

76

77

Step 7. Final edging. To put the final touch on the walk, use the edger all around the perimeter of the forms, cutting across the ridges made by the broom. This is a very easy operation, and you will have no trouble at all. See Fig. 78.

Step 8. Cutting the joints. If the concrete is not setting up too fast (as was the case on this day with the 40 degree temperature), you need not be in a hurry to cut joints. These joints were cut three hours after the concrete was poured. Check it frequently to make sure you are not waiting too long; especially on a hot day, the concrete may set faster. As long as you can still push the jointer's sharp edge into the surface, you are safe. It would probably be better to be a little early than late with this operation. Use your judgment. If you did not previously mark the forms for joints, you may lay out a tape, mark a mark in the surface at each point, and set your straight edge to mark the first joint. See Fig. 79. Do not lay the 2 × 4 straight edge flat. When you lean out on it in the middle, it may mark the concrete. One or two times across each joint should do it. Notice that the concrete is hard enough (three hours after delivery) to place the straight

edge on the surface (while putting in a corner joint) without support from the form. See Fig. 80. Keep in mind that the supportive hardness of concrete varies with temperature, humidity, wind etc.

Step 9. Curing process. It is important that your new concrete does not lose moisture too rapidly during the early stages of setting up. If it is very hot (80 or 90 degrees, for example) you can prevent this by shading it or covering

it with burlap or construction paper to keep off direct sunlight. If the temperature is fairly cool, or your walk is in the shade, or you pour later in the evening, simply water down your walk with the hose. Do this frequently for three days, and for a week if it remains hot. Wetting down the forms before pouring also helps.

Cold weather concrete work can be done if you follow some basic rules:

1. Thaw the ground with a torch before pouring, or put straw on it before it freezes. Remove the straw just before concreting.
2. Check to see if concrete company adds antifreeze material before delivery. They will usually do this without your instructions.
3. If you mix your own concrete, use about one pound of calcium chloride for each sack of cement.
4. The finished concrete should be covered with a thick layer of straw for four or five days to prevent freezing.

It is recommended that you choose good weather in the 50 to 70 degree temperature range for your first job. If this is not possible, you should take all precautions necessary to protect the concrete while you pour and cure the concrete properly. You may obtain additional information on winter concreting from the *American Concrete Institute, P. O. Box 4754, Redford Station, Detroit, Michigan 48219.*

POURING CONCRETE ON SLOPES

If your walk contains a section with a fairly steep grade, you may have to screed it several times before it will stay in place without having a tendency to run downhill. If it does slump after the first screeding, let it stiffen some before screeding again. When it is stiff enough, it will maintain a flat surface even on a slope.

FORMING AND POURING CURBS OR STEPS WITH YOUR SIDEWALK

Step 1. Building curb forms. If a curb is to be built separate from the sidewalk pouring, or on top of a previously poured walk, it can be done with a simple form of 2 × 6's held together as shown in Fig. 81. One side of the form is held in place by stakes in the ground while the other is tacked together with 6d nails toenailed into the 2 × 6. If possible, rods should be left protruding from the walk in order to tie into the curb. If the walk was poured without rods, drill into the walk every five feet, place a lead anchor and a ⅜ inch bolt sticking out 2 to 3 inches. See Fig. 82.

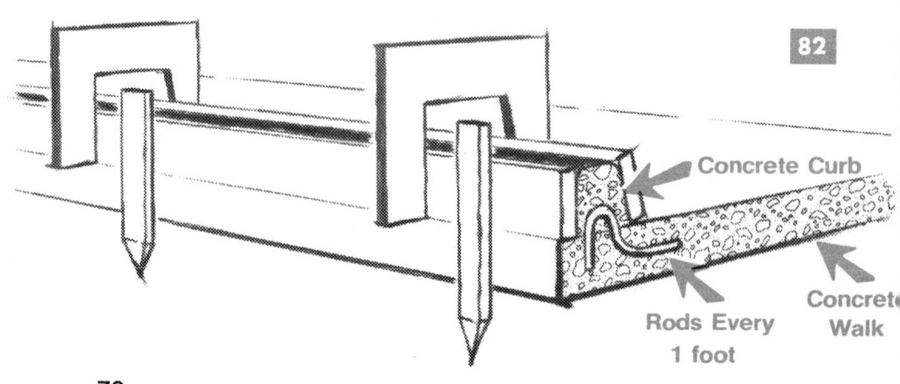

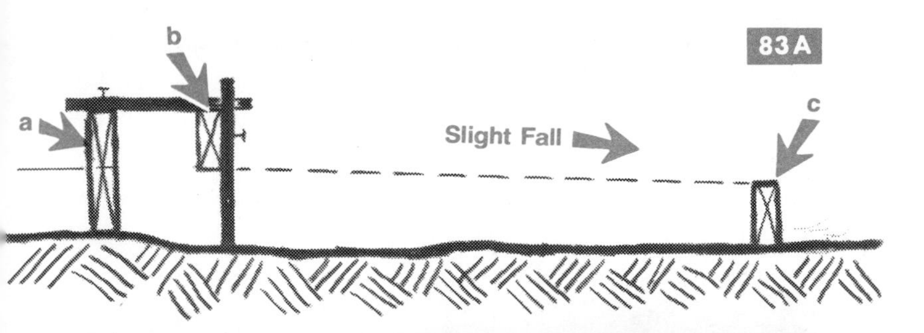

Step 1. For building a curb and walk form together. Set the forms the same as for a regular sidewalk, but use a 2 × 4 on the sidewalk side (b) and a 2 × 8 on the curb side (a). Mark the 2 × 4 width (about 3½ inches) from the top of the 2 × 8 and set the 2 × 8 so that this mark will be slightly higher than level with the 2 × 4 side (c). The slight amount of fall is for drainage. See Figs. 83A, B, and C.

Step 2. Pouring the curb. Fill the curb first. (See Fig. 83C). Let it be settling while you pour the walk portion. Puddle the concrete well in the curb to avoid honeycombs. When the concrete begins to stiffen, it will remain in the curb form without slumping.

84

85

Step 3. Finish the walk. Use a darby to float the fine particles to the top. See Fig. 84. Use the edger and jointer at the proper times (same as other regular walk) and then fill the low places in the curb and float to the top. Use an edging tool on the corners.

Step 4. Remove inside form. Three or four hours after pouring, when you can use the walk without damaging it, you can carefully remove the inside 2 × 4 form. If the stakes will not pull up out of the ground, drive them down as far as you can with a sledge hammer. Fill up the holes made by the stakes with some excess concrete that you have saved and kept pliable for this purpose. Refloat the damaged parts. A wet float can be used to touch up the inside of the curb where the form was removed. Hold the float flat on the curb side and lightly move in small circles. See Fig. 85. For this purpose a sponge float is best.

ADDING ONE OR TWO STEPS TO YOUR WALK

Step 1. The form. The lower walk can easily be lined up with the top walk by a line and level. Figure the height from the top of the lower walk to the top of the top walk (form). Your steps should be no higher than 8 inches, and each step should be the same height. The step treads should be about 10 inches wide. Attach the 2 × 8's to the plywood and dig out the hillside to set the form in place. Level the form and put a stake and dirt against the sides to hold it in place. See Fig. 86.

Step 2. Finishing the steps. Pour the steps first and let the concrete build up and stiffen. Keep adding concrete until the form is full and will not slump. Puddle the forms well by tapping with a hammer. Float and edge the step top as you did your walk, and put a broom finish or the same kind of finish on the step as you did the walk. The step should be given a fairly rough non-skid surface. Remove the forms later and finish the step faces.

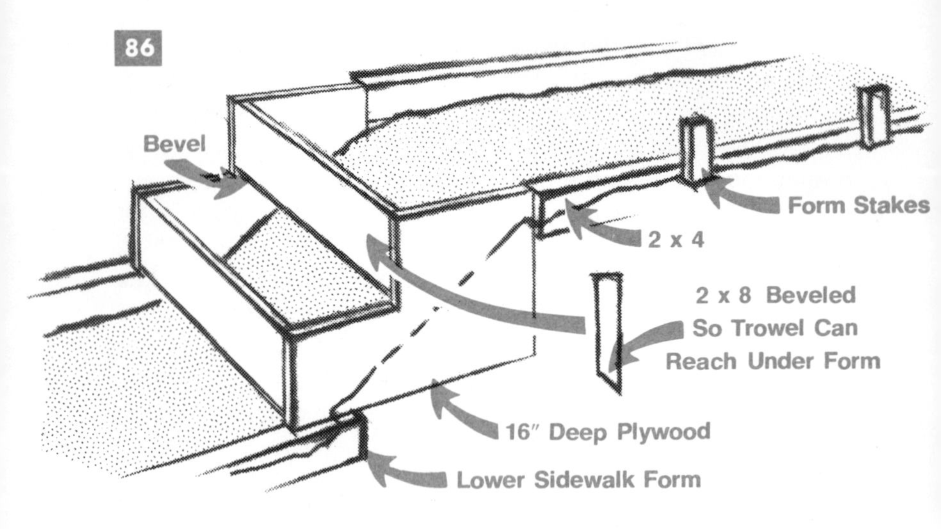

What to do about honeycombs

Step 1. Special mix. Use a special plastic-like mix of cement to fill in honeycombs. Finish with a brush. See Figs. 87 and 88.

Step 2. If you use other than special mix. Combine one half cement and one half fine sand well mixed with enough

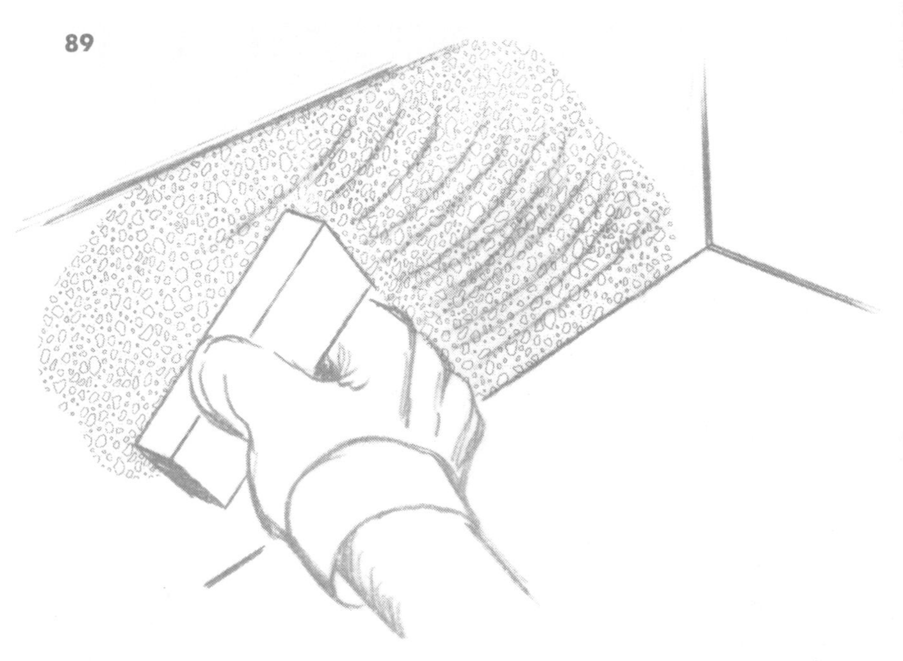
89

water to make a paste. Apply this paste to the step face with a small pointing trowel or broad knife. Rub out rough parts in the face with a carborundum stone. Keep the stone wet as you rub. See Fig. 89.

SPECIAL FINISHES

The form for this curved walk was made from 4 inch strips of masonite held in place with stakes at strategic positions. To achieve the exposed aggregate finish, the walk is first screeded and darbied as in pouring a normal walk. A colorful or uniform gravel (washed) is spread evenly over the entire surface. It is then patted with a darby or flat board until the aggregate is well imbedded in the concrete. When the concrete has set up enough to support your weight, you can use a magnesium float or darby to float the surface. Float it until the fine particles in the concrete completely surround the

gravel. Several hours later, hose and brush the top until the gravel particles are exposed. This will leave a beautiful and nonskid surface. See Fig. 90. Irregular stones can also make a very beautiful finish.

SECTION

SECTION

Porches, Driveways and Steps

Now that you have had the experience of building stepping stones and a sidewalk, you may wish to take on some more advanced jobs and also learn to repair broken concrete. The principles for forming and pouring bigger jobs are about the same with one exception; with a large amount of concrete pushing against a form, you will need more braces, stronger braces, and they must be closer together. Remember that one cubic yard of concrete weighs approxi-

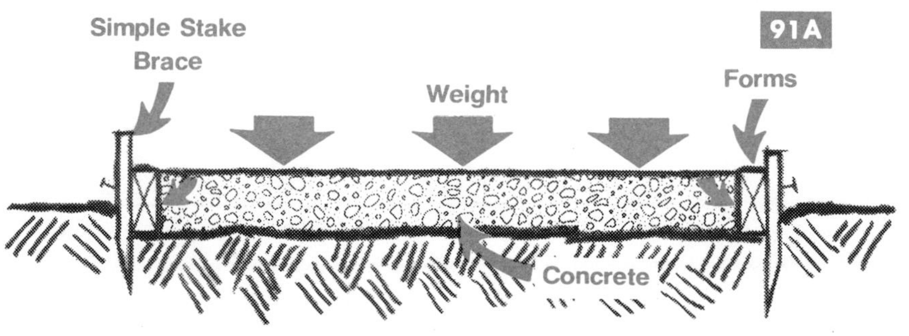

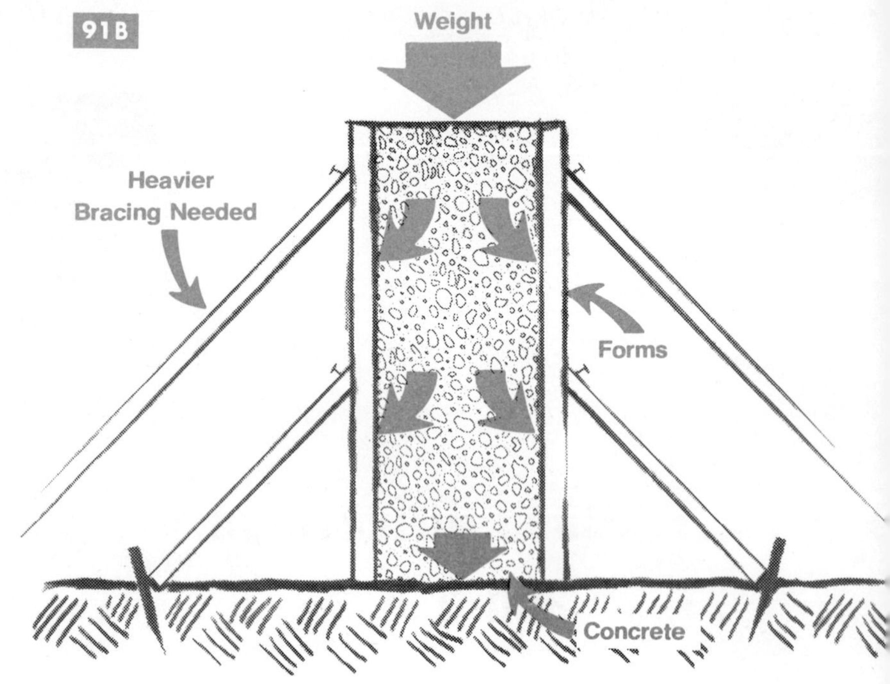

mately 4355 lbs. While it is in plastic form, gravity will not only pull it straight down but to the sides as well. In a flat pouring, (as shown in Fig. 91A) most of the weight is down so not much bracing is needed, but in an upright form, (as shown in Fig. 91B) the weight pushes to the sides.

HOW TO REPLACE A WOODEN PORCH WITH A NEW CONCRETE STRUCTURE

Step 1. Bracing the roof to remove posts. Most older wooden porches have at least three posts supporting the roof structure. If the roof is sound, you will only need to replace the wooden floor with concrete. Support the roof as shown in Fig. 92. Place three or four solid 2 × 8's (or 4 × 4's) at points supported by the present posts. Notch them slightly at the top. Make sure the beam is solid and not rotten where you are placing the support. Place the bottom of the support out about two feet from the prepared porch floor to allow room for concrete forms. Place the notched ends under the porch beam and drive the lower end in toward the porch with a sledge hammer, settling it into the ground and taking the pressure off of the posts. Have someone hold onto the old posts in case they are not fastened too securely. As you take the weight off, they could fall. When the weight is evenly distributed on the new braces, remove the old post supports. Drive 2 × 4 stakes into the ground in front of the braces to assure that they stay in place. Nail a large spike through the stake into the brace. You can strengthen your new supports further by nailing 2 × 4 braces between them if needed.

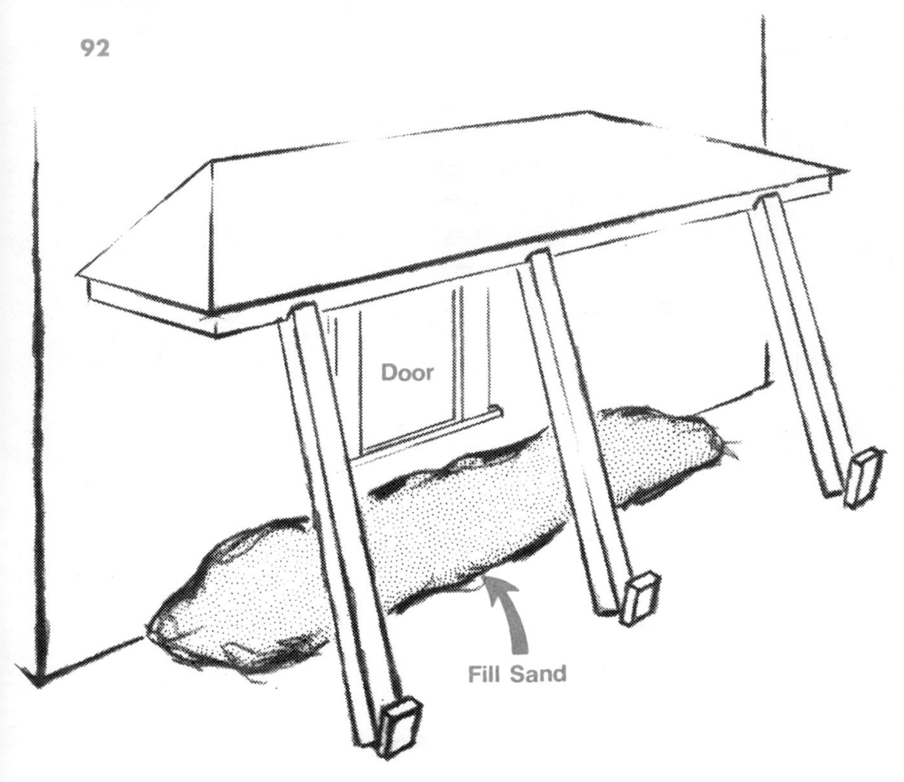

Step 2. Removing the wooden porch. Most likely, none of the wood in the old porch is worth saving. Exercise care in wrecking the old structure so that you do not injure yourself by falling through an opening. Remove the old lumber from the area where you are working so that someone does not step on a nail or the pile of lumber does not get in your way of digging the footings or building forms.

Two types of new porches

Your new porch will either be just a few inches off of the ground or over one foot off of the ground, depending on the height of the door threshold. If the new porch will be less than one foot off of the ground, proceed as follows:

Step 1. Low Porch. Set up a rectangular form the size of your new porch. You can use a 2 × 4 or a 2 × 6, or even a

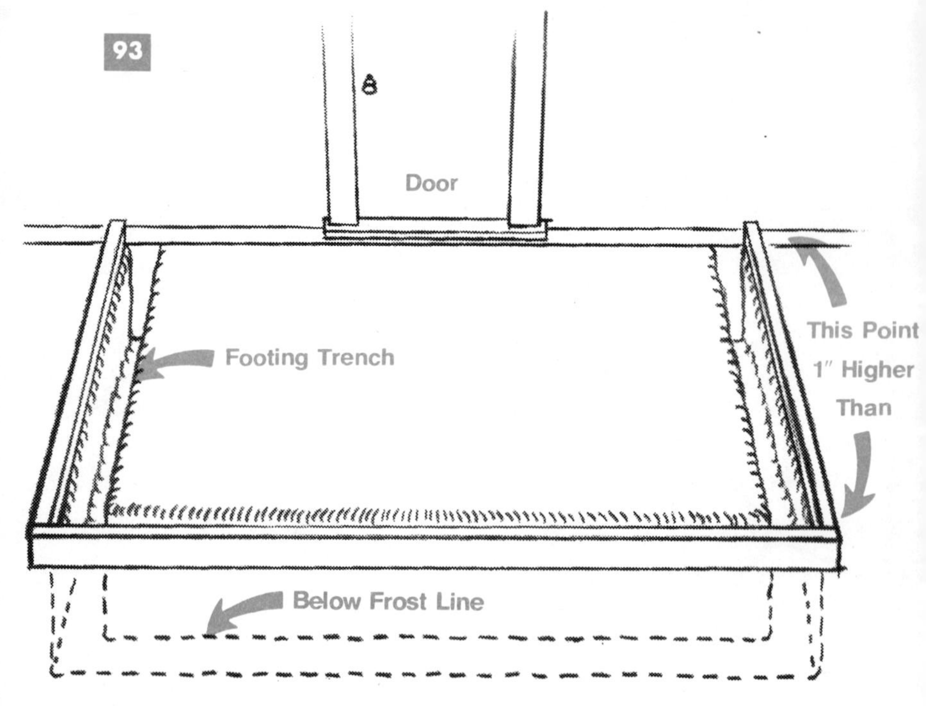

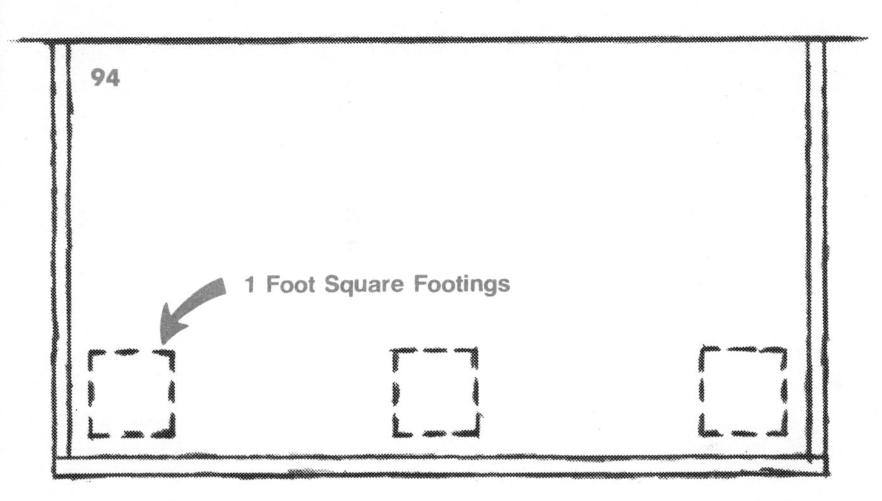

2 × 8 if needed. Use a form that will give you about one inch fall from the door to the front of the porch.

Step 2. Low Porch. Dig an eight inch trench around the inside perimeter of the form. (Fig. 93.) This is to provide a footing for a future brick wall on top of the slab. If you do not plan anything heavier than the present roof structure, you may want to dig three 12 × 12 inch footings below the point your new post will rest. In any case, dig below the frost line in your geographical location. See Fig. 94. In the temperate zone, this is about thirty inches. The local office of the Department of Conservation or any local concrete company can give you this information.

Step 1. High Porch. If your porch is going to be higher than one foot, proceed by digging a footing around the perimeter in which to pour concrete to support a concrete block foundation. Upon this foundation will rest the new concrete slab. Carefully read through the instructions before you start. You need to think about planning the right height of the footing so that the courses of block will come out right. You need to consider the size of the porch. If you plan well, you can make your concrete blocks work out to an even number. You will also need to consider drainage. Plan your foundation so that you will have about one inch of fall from

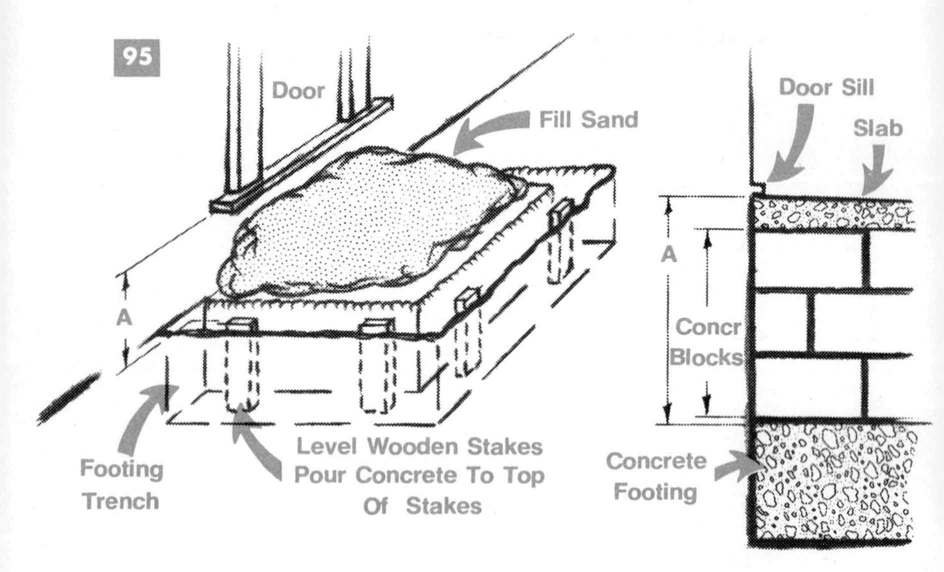

the bottom of the door sill to the front of the porch. Study the illustration in Fig. 95.

Step 2. High Porch. After you determine where the top of your footing should be, keeping in mind that each course of blocks is 8 inches high (including the mortar joint) and that you want the top of the top course to come up to within three inches of the door sill, set several small stakes in the trench to the height of the concrete footing. Put the first stake near the house, then use a long level (or a level on top of a straight edge) to level from the first stake to the next—on around the trench until you have a stake about every 4 feet. See Fig. 95.

Step 3. Pour concrete footing. Measure the total length of the trench. Multiply it times the depth and then times the width. If the measurement is in inches, i.e. 30 inches, use 2½ feet. For example, if the trench measure 6 feet out from the house and is 10 feet across the front, the length is 22 feet times the depth 2½ feet times ⅔ foot. This comes out to roughly 36 cubic feet. Since you know that there are 27 cubic feet to a cubic yard, you should order 1½ yards of concrete. Pour the concrete up to the stake tops and give it a rough float job. After the concrete hardens you are ready to lay the block foundation.

Step 4. Laying blocks. If you have figured correctly the even number of blocks it will take for each side of the porch, you should line them up along the trench, leaving ⅜ inch for every mortar joint. You should stretch a string from the house to the corner to mark where the block edge will be. If everything checks out, make a batch of mortar from ½ sand and ½ mortar. Put the mortar on a board near your work. With a trowel full of mortar, start laying out two lines of mortar on the footing. The main edges of the block will set in these tracks. Do not fill the entire footing with mortar as it will be wasted. Again, it will help you to go to a construction site to watch block layers work. The correct technique to lay down the mortar is to make a swinging motion above and parallel to the footing, laying out the whole trowel full in one sweep. Reload the trowel and continue the whole side before laying any blocks. Always remember to wet each block before applying mortar. If blocks are not damp they will absorb moisture from the mortar, which will weaken the mortar joint.

Now with the blocks on end, hold a partial trowel of mortar, as shown in Fig. 96, tilt the outside of the trowel up and make a swipe down "buttering" the block on one end.

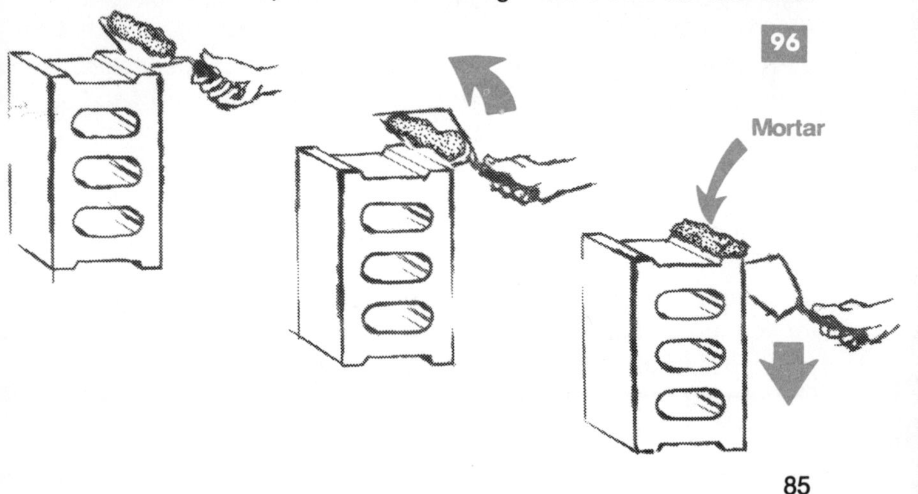

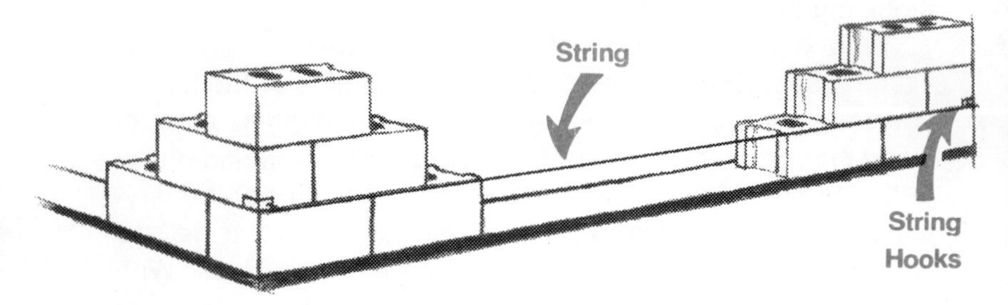

Pick up the block by the narrow webs and lay the "fat" side down. Adjust it until the edge is along the string.

Lay one side of the foundation to the corner, using a corner block at the outside end. Lay out the other side the same way, using string and a level on top. Now lay out the front course. On the second course, lay up the two corners to the correct height; stretch a string for the edge and fill in the middle. See Fig. 97. You will have to use a half block occasionally. Some blocks are made with a double web in the center. These can be broken easily with a hammer and chisel. As you lay the blocks, keep checking with a level to see how straight you are going. You can use your level as a straight edge to push against the side of the blocks to line them up. You can also "eyeball" your work. (Stand back and sight it for straightness).

Step 5. Filling with sand. You can start filling up the inside of your foundation right away with sand. Do not tamp sand against the freshly laid blocks for a day or two. Even then, you should brace your new wall from the outside until the mortar hardens sufficiently. See Figs. 98A and B. Fill the holes in the top of the blocks with broken blocks, newspapers and sand. If this is not done, you will need more concrete to fill up these holes. Now you are ready to build the form for your porch floor.

Step 6. Building the form. Using long straight 2 × 8's nailed to 2 × 4's, as shown in Fig. 99, set the front form first so that the top edge is 1 inch below the point where the

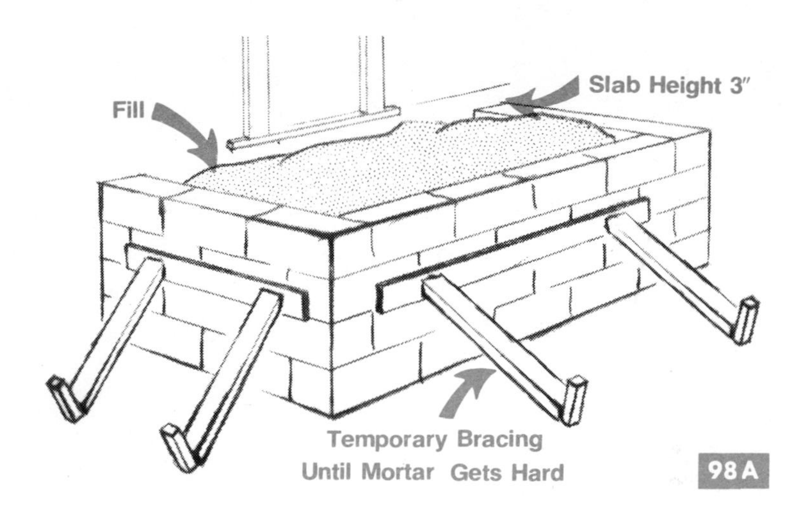
Fill
Slab Height 3″
Temporary Bracing
Until Mortar Gets Hard
98A

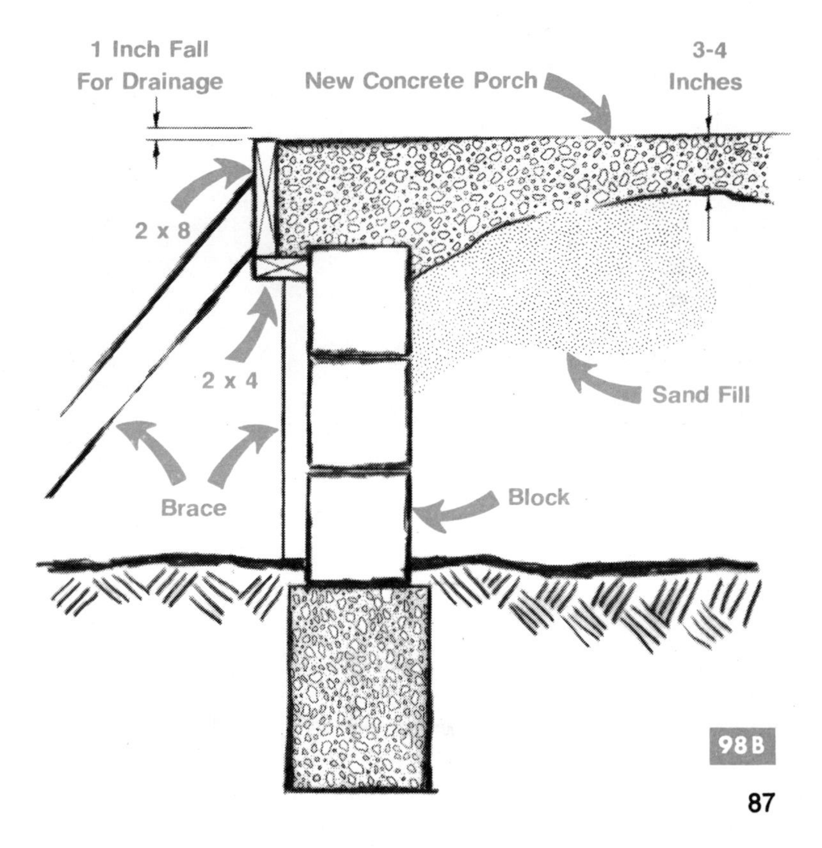
1 Inch Fall
For Drainage
New Concrete Porch
3-4
Inches
2 x 8
2 x 4
Sand Fill
Brace
Block
98B

concrete will come in contact with the door sill. You can check this at various places by establishing a chalk line on the house. Either drive a small nail into the chalk line at several points or drive a small stake down to that height. Take a straight 2 × 4 reaching from the house to the front form, place a level on top of it and adjust the form accordingly. The form can be held to the correct height by short 2 × 4 driven into the ground under the form, and braced with an angle brace at the top edge. The side forms will help hold the front in place. Use wire mesh to reinforce the floor.

Step 7. Pouring concrete. The technique will be about the same for both high and low type porches. If there is a footing trench to fill, pour the concrete in this area first and spread out the concrete evenly against the form as you go, not putting a strain on one part of the form at one time. Fill up to the top of the forms and screed it off as you did the sidewalk. See Fig. 99A and B.

If you need a guide for your screed next to the house, set small stakes (metal or wood) about one foot out along a line parallel to the house as shown in Fig. 100. Drive these stakes down to a level to hold the pipe up to the correct level. This will make a nice surface to drag the screed. After you hand float next to the house, remove the pipe, drive down the stakes. Fill in the holes with excess concrete and use the bull float across the entire top.

Although regular concrete is all right to use, it is recommended that you ask for air-entrained concrete. It is the same price and it contains less water per cubic yard. It is also easier to work than regular concrete. Be sure to puddle well (tap on the forms with a hammer) so that you will not have honeycombs on the porch edges. Another good thing to do

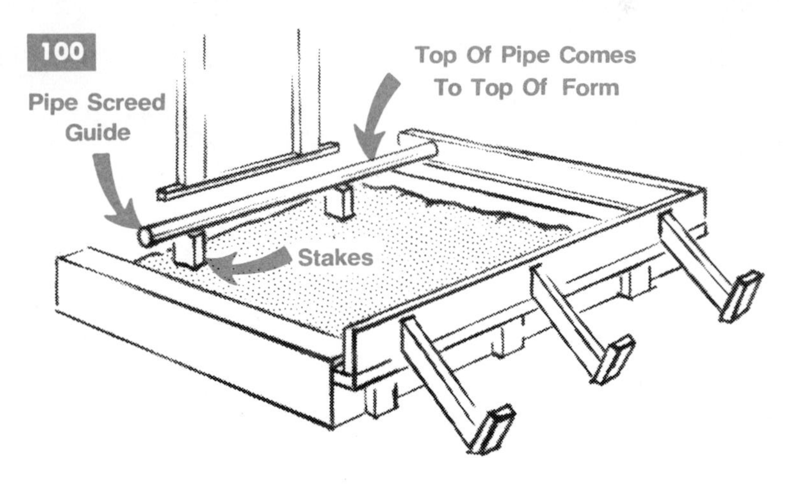

is use paraffin oil on the inside of the forms or at least wet them down before pouring. If it is a hot day, wet the sand down also, or your surface may dry out too fast. Use the finishing techniques described earlier in this book. Float with an aluminum float; then use a steel trowel. A trowelled surface or a swirl finish looks better than a broom finish on a porch. For traction, a swirl finish is recommended. See section on stepping stones.

Step 8. Removing the forms. After the porch surface is hard enough to support knee boards to trowel, you can carefully remove the side forms. Using a sponge float and a mixture of ½ sand and ½ cement mixed into a paste, float the sides until you get a suitable surface. Keep the concrete watered down for a few days until it cures. You may have to put up barricades to keep people from walking on the new surface. The edges are very vulnerable to breaking or crumbling and should cure for at least 2 days.

Step 9. Steps. A low porch (of course) needs no steps, but for a high porch you will need to build step forms and pour one, two or three steps as shown in Fig. 101. Finish the same way you did the porch. You may also pour a front walk at the same time.

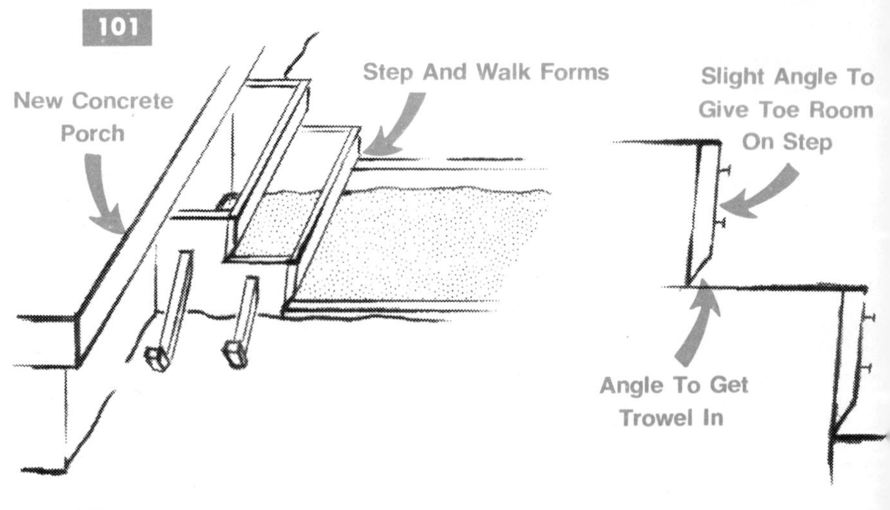

REPAIRING OLD STEPS

While it is better to build new steps, at times it is more economical to repair the old steps until such a time when you can build new ones. See Fig. 102. You can make the older steps look fairly good by following these instructions:

Step 1. Breaking up old steps. With a sledge hammer, break up the damaged portion of the step. If there are any hairline cracks visible, it is almost certain to be a weak area. Clean all loose debris from the broken area. See Fig. 103.

102

103

104

Step 2. Building a form. Use 2 inch lumber to hold the new concrete in place. The form does not have to be "pretty" to do the job or to be the same height as the old step. You can use a straight edge inside the form to screed the concrete to the right height. See Fig. 104.

105

Step 3. Placing the new concrete. Mix up a bag of premixed concrete (see stepping stones). Tamp the mixture well into the corners and along the form. Tap the front with a hammer to avoid honeycombs.

Step 4. Finish the concrete. Using the edger, float and trowel, as you did on the stepping stones, finish the surface to match any existing concrete work. Let the steps harden for a couple of hours and remove the forms. Then finish the front of the steps with grout and a sponge float as you did the porch front. See Fig. 105.

POURING DRIVEWAYS

Once you have the experience of building a concrete walk, you will find it easier to pour a concrete driveway. If you have a very wide drive (ten feet or so) and very long (over 30 feet), you may find it easier to pour it in two sections. At any rate, you will probably do it for one third the cost if you do the work yourself.

Step 1. Building the forms. The drive shown here is about 25 feet from the street to the garage and built at a slightly steep grade. The forms are made very simply with 2 × 4's, held by stakes as in sidewalk building, and supported with loose sod shoveled against the 2 × 4 forms.

Step 2. Pouring the concrete. The driveway should be graded out at least four inches so that the concrete is close to six inches deep. However, 4 inches will hold most cars. It is also better to reinforce the concrete with wire mesh and to provide expansion joints between the garage and the drive. Order not over a number 6 slump and not under 3500 to 4000 pound strength concrete. The slump refers to the wetness and the strength to the formula of aggregate mixture ratio. If you are pouring on a very steep slope, it might be well to order a number 4 slump. This stiffer mixture would not tend to run downhill like a wetter mixture. The disadvantage of a lower slump mixture is that it is harder

to work. If you have several helpers, this will overcome that disadvantage.

Pour the concrete into the forms as you did on the walk. See Fig. 106. You will have to use a longer screed, which of course will be a little harder to pull, but it is basically the same technique. See Fig. 107.

Step 3. Finishing the concrete. Immediately after screeding, employ the large aluminum bull float with a longer handle. See Fig. 108. There are several length of handles to screw into the float to use on a variety of size jobs. You will find that the bull float is one of the best work savers ever intended for concrete work, not counting the motor driven troweler of

course. You should use an aluminum or magnesium darby to go over the surface when it begins to harden. You should use the edger along the sides and cut joints in strategic places as you did on the walk. First, you should put a broom finish on the driveway, with the lines running perpendicular to the tires that will run over it. Cure the concrete the same as the walk. Remember that you should not drive on it for over a week. It will reach maximum strength after 28 days.

Step 4. Removing the forms. Place the edge of a pick under the forms and use the handle for leverage to lift the 2 × 4's up and away from the concrete. Grade the dirt back around the new concrete.